YOUR BODY CLOCK

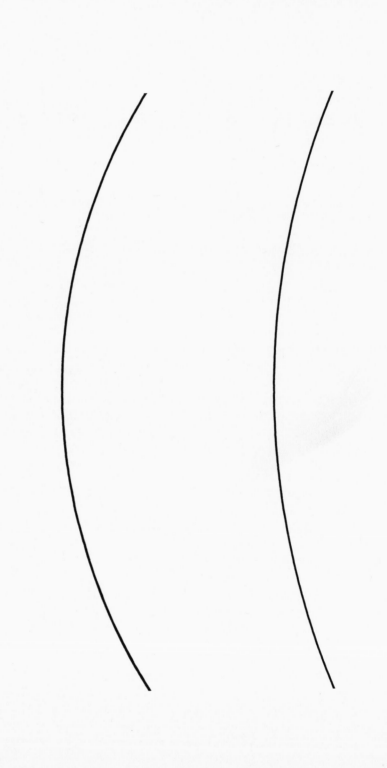

YOUR
BODY
CLOCK

Hubertus Strughold

Foreword by Charles H. Roadman

Charles Scribner's Sons / New York

To my wife,
Mary

Acknowledgments

I would like to express my appreciation to Thurman A. Glasgow for editorial assistance; and to Betty DePlachett, Gertrude B. Stephens, and Claudine Skeats for secretarial work.

Contents

Illustrations

Foreword

DOCTOR HUBERTUS STRUGHOLD is often referred to as the Father of Space Medicine. Through his many years of devoted effort in the fields of aviation and space medicine, he has become recognized as an authority on many subjects. One of these is the physiological clock (body clock), which regulates our sleep and activity phases within the cycle of approximately 24 hours.

Doctor Strughold's first paper on the subject—"Physiological Day/Night Cycle after Global Flights"—was published in October 1952. His interest had been stimulated two years before, when he flew for the first time between Washington, D.C., and Frankfurt, Germany. After two months in Germany, he flew back to San Antonio, Texas, where he was Chief of the Department of Space Medicine at the Air Force School of Aviation

Medicine. His observations concerning his body's adjustments to new time zones after these two transatlantic trips stimulated him to look deeper into the problem.

Until air travel became commonplace, the problems encountered in rapidly spanning many time zones were of little importance. Now that many more travelers are covering greater distances in shorter periods of time, closer attention will have to be given to the effects on the body clock. This book is intended to help travelers who cross many time zones in a short time to understand their body responses and to suggest what they can do to minimize their discomfort.

The author also discusses the sleep and wakefulness time patterns of the astronauts in near-earth orbit and during flights to the moon. His book contains many physiological details of the circadian cycle or rhythm of the human body, and a number of interesting facts about sleep which are not generally known.

Charles H. Roadman
Major General, U.S. Air Force (retired)

YOUR BODY CLOCK

Introduction

During the past fifty years, enormous progress has been made in medical and biological science; in fact, man has entered the golden age in medicine and biology. To a great extent, this advance has been achieved because of the close cooperation between the biomedical sciences and physical sciences such as physics, chemistry, and electronics. One of the fruits of this cooperative effort is a better understanding of the miraculous timing device in the human body which regulates the sleep and wakefulness cycle: the biological clock or body clock.

From the time of the caveman through the horse-and-buggy era of transportation, the time for activity and for sleep was natural and uncomplicated. However, in the modern era of industrialization, electric illumination, worldwide audio and video telecommunication, and motorized transportation, particularly by jet and rocket propulsion, the simplicity of time has turned into complexity. As a result, man has become more time-conscious; therefore, he must have a better understanding of his internal clock. In anticipation of this development, researchers in so-called biorhythmology have focused more and more attention upon the physiological clock in relation to rapid and drastic changes of the environment.

1

Rhythm—
The Rule of
the Universe

A FUNDAMENTAL PHENOMENON in the universe is the occurrence of cycles. Cycles, or rhythms, can be defined as repetitions at regular time intervals of situations, events, and levels of activity. Almost all processes in the universe are cyclic. This phenomenon is found on a gigantic scale in the vast dimensions of the universe, on a miniature scale in the microscopic world of the atom, and in the processes of life on earth.

Beginning with the physical nonliving world, the galaxies rotate around their centers. It takes the galaxy to to which the sun belongs about 200 million earth years to make such a round. This means that every 200 million years the sun returns to the same place in space. If 5 billion years is assumed as the present age of the solar system, it is now in the 250th galactic year.

The sun itself rotates around its axis within 25 earth days. Furthermore, it shows an 11-year cycle of increased activity, manifested at its peak by frequent solar flares. These explosive eruptions, accompanied by the appearance of dark spots on the sun's surface, produce violent solar storms during which large masses of atomic particles, such as protons and electrons, are ejected and blown deep into space. When a stream of these electrically charged particles, called solar plasma wind, hits the earth, it has a marked effect on the ionosphere, or upper atmosphere, causing a black-out of long-distance radio communications which may last several days. The stream also affects the lower atmospheric regions by triggering rainy weather.

The planets revolve around the sun in more or less circular orbits to form the cycle referred to as a planetary year. The terrestrial (earth) year lasts about 365 days. Within this annual cycle there is a sequence of seasons: spring, summer, fall, and winter. The seasons are distinguished by differences in light intensity, temperature, and humidity.

Comets moving in elliptic orbits become visible to us at regular intervals; Halley's comet, for example, appears every 75 years.

Frequently the earth enters a meteor stream formed by the remnants of a disintegrated comet, and at such times the spectacle of a meteor shower can be observed. This happens regularly on certain dates in the year.

The moon, during its revolution around the earth, displays an illumination cycle—from new moon to full moon and back to new moon—of about 27 days. This is a result of the relationship of the moon to the earth during the movements around the sun.

The earth's rotation around its axis in the radiation

field of the sun leads to a light/dark cycle combined with a temperature cycle. This is a day/night cycle lasting 24 hours.

In the course of the earth's rotation, the moon's gravitational pull produces high tides on the seashores every 12.5 hours, with low tides in between. The levels of the tides are modified by the gravitational force of the sun. So much for the cycles in the realm of the universe, the macrocosmos.

As for cycles in the microcosmos or microscopic world, just as the planets revolve around the sun, the electrons move with tremendous speed in orbits around the nucleus of the atom.

Rhythms or cycles are characteristic also of the living world as it exists on earth. Practically all life processes show cycles in their levels of activity. These cycles can be categorized into two basic types. The first is called an *exogen* biological rhythm. It is caused by, and is dependent on, external environmental cycles. The second type consists of the *endogen* biological rhythms, most of which originated independently as an inherent property of life. In the course of evolution, they have become influenced by and synchronized with external, physical, and rhythmically recurring environmental factors of the nonliving world. Of the physical cycles of the universe, only those within the solar system have an effect on these two types of biological rhythms.

The 11-year cycle of maximal solar activity is the one of longest duration. It not only affects the earth's atmosphere but has a definite influence on plant life. For example, the yearly growth rings in the trunks of trees show a pattern of increased size every eleventh year, as a result of the more frequent rain brought on by the increased solar activity.

The annual cycle of seasons is reflected in the growth rate, color, blossoming time, and other aspects of all green plants. These rhythms in the plant kingdom are completely dependent on the environmental cycles and are therefore exogen in nature. If, as some scientists believe, there is plant life in the dark areas on the planet Mars, then the seasonal color changes from dark to bluish green, yellow gold, brown, and back to dark would be an extraterrestrial example of this type of biorhythm. In the animal kingdom, a remarkable seasonal cycle is the regular north and south flight of migrating birds, such as cranes, geese, and swallows. Another is the hibernation of some animals in response to extremely low temperatures.

The next shorter cycle, the new moon/full moon cycle, has a distinct effect on the life of nocturnal animals. Records show that, during full moon, there are a hundred times more female mosquitoes in the air chasing male mosquitoes than during the time of the new moon. Some humans claim that they do not sleep well during the full-moon period.

The most impressive and most familiar biological rhythm is the one associated with the earth's axial rotation in the powerful electromagnetic radiation field of the sun, the light and darkness cycle or day/night cycle. This biorhythm of a period of 24 hours is called the *circadian cycle;* the name is derived from the Latin words *circa* and *dies,* meaning approximately one day. The most conspicuous physiological features of this circadian rhythm are the phases of sleep, or rest, and wakefulness. This is the rhythm of man's internal physiological clock or body clock. Responses to the day/night cycle occur among animals and in numerous plants as well. The rotating earth can be said to be the timekeeper for all its inhabitants.

Another and shorter environmental cycle is the 6-hour alternation of high tide and low tide, with which the activity/rest rhythm of seashore animals is synchronized. In this case the moon is the timekeeper. These are only a few of the multitude of rhythms in the living world on earth or in its biosphere.

The human mind, too, is attracted by rhythms. This psychophysiological attitude is best demonstrated by man's love for rhythms in poetry, music, and dance. Much of man's entertainment and relaxation revolve around rhythm in some form.

The scientific interest in biological rhythms has increased tremendously during the past two decades. It has led to the development of a new branch of biology —biorhythmology or chronobiology—and to the founding of a scientific organization, the International Society for Biological Rhythms. Both the science and the organization are concerned with the whole spectrum of rhythms found in plants, animals, and man.

2

Structure and Functions of the Physiological Clock

In order to understand the functions of the body clock, one must become familiar with the nature of sleep and wakefulness. This, in turn, requires an analysis of the fundamental structure of our body's clock —its anatomy, its physiology, and its development through the stages of human life.

Levels of Wakefulness and Sleep

During the state of wakefulness, man is conscious of the outside environment as well as of processes going on within his body. This awareness is possible because of the exteroceptors and the interoceptors, or proprioceptors, of the various sense organs. The exteroceptors include those for vision, hearing, smell, taste, touch, and temperature perception. The interoceptors provide awareness of pain, tension, pressure, and hunger. There are various degrees of wakefulness, expressed by such terms as alertness, attentiveness, and vigilance. Mental and motor activities comprise a large part of the state of wakefulness. Drowsiness can be regarded as the intermediate phase between wakefulness and sleep.

During sleep man becomes less and less aware of the outside world as he passes through the various stages in depth of sleep. Light sleep is the first stage; medium or moderate sleep the second; near-deep the third; and deep sleep the fourth. Usually, man falls into deep sleep within five or ten minutes after retiring, and this stage lasts about two or three hours. Then follow fluctuations of light and medium sleep, with occasional short dips into deep sleep or back to wakefulness. Dreams, usually accompanied by rapid eye movements, occur during light and medium sleep. This rapid eye movement is sometimes referred to as REM sleep. The phases of dream activity, with or without REM, are now called paradoxical sleep.

During deep sleep there is little or no response to noise, but during light and medium sleep people wake easily at the slightest sound, especially if it has either a pleasant or a frightening connotation for the sleeper. It is a well-known fact that a mother, although able to sleep soundly through a thunder and lightning storm, can be awakened by the slightest voice noise of her baby. One mother reported that the whimpering of her dog had the same effect because of its similarity to her baby's cry. This is called a "wake island" in the brain, a fascinating psychophysiological phenomenon for which there is at present no neurophysiological explanation.

26

In addition to the 6 to 8 hours of night-long sleep, many people enjoy a short afternoon nap. Furthermore, after a night with insufficient sleep, a person occasionally experiences sleep seizures that last only a few seconds. These seizures, which are called microsleep, are frequently the cause of automobile accidents. Since microsleep is used to designate these brief sleep seizures, an appropriate term for the afternoon nap lasting 30 to 40 minutes would be "minisleep."

Anatomy and Physiology of the Clock

All the stages in wakefulness and sleep are characterized by distinct patterns of the electric activity of the brain. These brain waves, as they are called, can be measured by an electroencephalograph.

During the state of wakefulness, the electroencephalogram shows a dominant frequency of brain waves of about nine to thirteen oscillations per second and of several millivolts—the so-called alpha waves.

With the onset of light sleep, this rhythm changes slightly into slower, larger waves. During medium sleep, short bursts of waves called sleep spindles appear. Dreams occur during light and medium sleep *after* the deep-sleep period. During this dreaming phase, the electric waves show a higher frequency. A nightmare can cause bursts of more frequent and violent oscillations. During near-deep and deep sleep the frequency decreases to two or three oscillations per second, with increased voltage; these are called delta waves.

As mentioned earlier, sleep and wakefulness, the latter characterized by mental and sensorimotor activity, are the most conspicuous signs of the circadian rhythm. But there are many more rhythmic changes behind

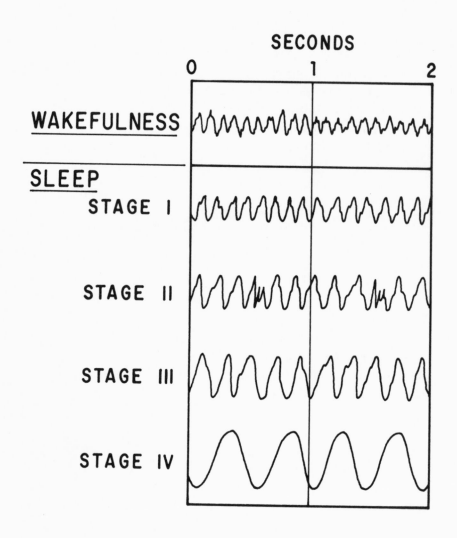

Electroencephalograms of human brain waves during restful wakefulness and the four stages of sleep

the visible scene. These are found in the activities of nearly all body organs. Their specific functions, or the processes associated with them, can be recorded by special electrical recording or biochemical methods.

During sleep, there is considerable relaxation of the tone of the voluntary motor muscles, and their activity practically disappears except for some dozen reflex position movements of the body. These are stimulated by the pressoreceptors, or pressure points, of the skin regions under pressure. The movements can be recorded by a device called a hypnograph, or actograph, attached to the springs of the mattress.

This inactivity and relaxation of muscle tone results in a lower rate of metabolism. Under conditions of light workload and normal nutrition, the body requires a total of about 500 liters of oxygen during a 24-hour period. Of this amount, only 20 percent is consumed during the hours of nightly sleep. About 80 Calories per hour are produced by the human body during sleep as compared with 180 Calories per hour during the period of daytime activity.

Consequently, during sleep, respiration slows down from the daytime ratio of about 16 breaths per minute to 12, and the heart rate from about 70 to 60 beats per minute. Blood pressure is lower as well.

Recently it has been found that the sensitivity of the taste nerves in the tongue is at its maximum in the early evening. This has been regarded as the reason why most people have their main meal at that time. As a consequence, the secretory activity and the peristaltic and antiperistaltic movements of the stomach and intestines increase during the night.

At the end of this anatomical line of nutrition and digestion, the elimination of waste products via the rectum (normally a rather punctual timekeeper) and the urinary system takes place predominantly during the day, permitting sound and undisturbed sleep at night. Urine analysis has revealed periodic fluctuations in the concentration of salt and other chemicals that may be coupled with food and water intake. The blood is a kind of mirror, reflecting in its cellular and chemical constituents a picture of the body's overall activities in the form of day/night variations.

According to recent analytical studies, the blood plasma and the urine have proved to be important sources of knowledge concerning the hormone production of the endocrine glands. This has specific significance, since hormones, such as adrenalin, cortisone, thyroxin and others, play an important role in the regulation of the circadian rhythm. Their production rate shows a distinct circadian time pattern, manifested in maximum, minimum, and in-between fluctuations corresponding to their specific physiological functions. For example, the amount of adrenalin in the blood plasma fluctuates. This hormone, produced by the medulla of the adrenal gland, has a stimulating effect on the heart and skeletal muscles and increases the rate of carbohydrate metabolism. After a midnight low, the adrenalin content of the blood plasma reaches a peak between 8 and 9 a.m. and another around 2 p.m. These peaks may be related to the time of food intake.

The timing of the activities of the various endocrine glands is coordinated by several hormones of the pituitary gland. This "master gland" additionally produces a growth hormone, called somatotropin, at a higher rate during rest and sleep; as a consequence the multiplication rate of the cells of the bones and muscles is higher under these conditions. The pituitary gland is located at the base of the brain, well protected from outside injuries. It receives its orders, so to speak, via nerve fibers from the hypothalamus of the interbrain. This central station of the autonomic nervous system, with its sympathetic and parasympathetic divisions, controls all activities relating to growth and nutrition. During sleep, the parasympathetic system is dominant over the sympathetic; during the state of wakefulness, the situation is reversed. This explains the day/night differences in the activities of the respiratory, cardiovascular, digestive, and elimination systems.

All these periodic variations of the body's activities—harmoniously integrated into a functional system, with the hypothalamus as the coordinating center and the hormones as the intermediary chemical agents, but strongly influenced by the mental activities of the cerebral cortex—repeat themselves with clocklike regularity within the temporal frame of 24 hours. Hence the system has been termed the physiological or biological clock. It is a neurohormonal system, or in a broader sense a psychoneurohormonal system, which affects all of man's activities, including sex. According to a report in the chronobiological literature, the male sex is most sensitive to female charm between 3 a.m. and the late morning hours.

The best-known and most readily measurable indicator of the physiological clock is the body's temperature, with a high around 5 p.m. and a low between 4 and 5 a.m. But the system is actually composed of various subsystems with their own individual clock indicators. The clock system as a whole is predominantly a one-wave circadian device, but the subsystems show additional fluctuations in their activities within the 24-hour period in order to meet the needs of the whole body at any given time.

This multiplicity within the physiological clockwork has some implications for medicine. The effectiveness of drugs, for instance, varies at different hours of the day/night period. Franz Halberg, professor of physiology at the University of Minnesota, has made intensive studies of this problem of timing medical therapy to determine the correct dosage at the right time. Furthermore, according to Dr. R. F. Fitch, Chief of Internal Medicine at Wilford Hall USAF Medical Center, San Antonio, Texas, the administration of hormone-containing drugs should simulate the natural circadian production pattern of the hormones, to avoid disturbing their role in running the physiological clock. In this connection, Dr. J. R. Robison, Chief of Urological Service at the same hospital, has reported that a transplanted kidney gets in tune with the activity cycle of the other kidney within about a year. An artificial pacemaker for a defective heart is automatically regulated to give about ten fewer electric impulses during the night than during the day. Finally, certain symptoms, such as pain connected with numerous diseases, are more pronounced during certain hours of the day.

The activity of the body's internal clock is primarily governed by the light of the external environment. The natural stimulus, or time signal, for the onset of wakefulness and of sleep is therefore the change from darkness to light at sunrise and back to darkness at sunset. Dawn and dusk are the *Zeitgeber*, or time cues, for entraining the phases of sleep and wakefulness. This applies to both diurnal, or light-active, and nocturnal, or dark-active, living creatures. Color vision, which requires a certain light intensity, may play a stimulating role in the case of light-actives, except for those that are color blind. Certainly birds start their morning concert as soon as their surrounding world appears in color.

The significance of the change in illumination as a time cue was demonstrated during the total solar eclipse on March 7, 1970, which caused diurnal animals to go to sleep and nocturnal animals to wake up. In Florida, chickens went to their roosts and mosquitoes attacked their human prey. The solar eclipse was a kind of time deceiver that upset the body clocks of animals such as birds and insects.

But the change in illumination from dark to light and back to dark is not the only time cue for entraining rest or sleep and wakefulness; changes in temperature and humidity also play a role, especially in the case of plants and cold-blooded animals such as insects, fishes, amphibians, and reptiles. Other pertinent factors are security and hunger, which are usually obvious in warm-blooded animals.

Sleep Patterns of Some Animals

Although a general comparative physiological consideration of the biological clock is very revealing, a survey of certain well-known representatives of higher animals provides insight into the historical development, or phylogeny, of the biological clock.

Security dominates to a high degree the circadian rhythm of many wild and domestic animals which are constantly in danger of being attacked by predators. These animals are light sleepers, and their biological clocks are multicycled, or polycyclic. Among such prey animals are the rabbit, sheep, goat, antelope, and baboon. However, certain other prey animals are able to protect themselves in such a way that they are able to sleep soundly. The mole and ground squirrel, which live mostly in extensive networks of burrows, are good sleepers for periods of eight to fourteen hours per day. Others, such as the macaque monkey and the chimpanzee, find their protection by sleeping in the tops of tall trees. But the good sleepers among prey animals are the exceptions.

In contrast, the predators, such as the lion, tiger, leopard, wolf, fox, dog, and cat, are essentially good and long sleepers.

All these predators are carnivorous or omnivorous. Most of the prey animals are herbivorous; their feeding habits and method of stomach digestion, manifested in rumination, make their sleep pattern polycyclic. Sheep and pigs, for example, are mostly awake between the hours of 6 a.m. and 6 p.m. with frequent naps in between. The animal whose sleep patterns are best known is the horse, thanks to the detailed studies of 600 cavalry horses made by Dr. T. Steinhart, a veterinarian in Germany. As reported in his book, *The Sleep of the Horse* (1937), horses sleep essentially in periods of two hours at a time. The longest waking period is from 3 to 8 p.m.

Amazing are the recorded observations of this "horse watcher" concerning the four body positions during various stages of the animal's sleep. During the period of light sleep, a horse stands dozing with its head held free. During periods of moderate sleep, the horse rests its chin or head on a fence rail, a manger, or the back of another horse. During periods of deep sleep, it lies

(1) Light sleep

(2) Moderate sleep

(3) Near-deep sleep

(4) Deep sleep

Body positions of a horse during four stages of sleep

on its belly or, more often, on its side. In this detailed study, the differences in depth of sleep were measured in various ways, ranging from the application of stimuli applied to the sense of smell or sight, to the firing of a rifle shot from varying distances.

Development of the Human Physiological Clock

The development of the human circadian rhythm with age (its ontogeny) is of special interest. Babies, after emerging from the lightless uterus, show more than half a dozen alternations of sleep and wakefulness in the first two or three months of life. They are poly-cyclic or minicyclic. For them, the time signal for awak-ening is not the rising sun but the more frequently rising appetite, indicated by the periodic occurrence of stomach contractions. They are essentially food-active, whether it is dark or light. Their stomach clocks deter-mine their behavior. Their sleep/wakefulness time ratio is 2 to 1; that is, they are asleep during two-thirds of the 24-hour period. During this early part of their lifetime, babies behave like both nocturnal and diurnal birds, such as the owl and the starling. After three months, however, they become increasingly curious about the surrounding world and more and more light-active. They gradually consolidate the multitude of short phases of sleep and wakefulness into one long nightly sleep, with wakefulness predominating in the daytime. Now they behave essentially like starlings. The head clock is becoming dominant over the stomach clock. When, after about the ninth month, babies start to walk, they are, by and large, monocyclic. Their body clocks are fully developed, as can be proven by the main indicator, the temperature curve. But young children still need more sleep than adults do—about 10 hours during the night with several short naps during the day. Only after children have reached their teens is their sleep/wakefulness-time ratio 1 to 2, equivalent to that of the adult. Although this developmental pattern shows great individual variations, it provides the basic developmental characteristics of the sleep profile for comparison with that of the adult.

Sleep Requirements of Adults

The normal physiological time requirement for sleep, or the normal dosage for an adult, is about 7 hours in every 24, give or take an hour. This period, which is normal for people who are not under too much social or professional pressure, includes the night-long sleep and a half-hour catnap during the day. People over 60 need less sleep, a total of 5 to 6 hours usually being sufficient.

To measure sleep, one must consider both its duration and its depth. Most people, on retiring for the night, fall immediately into a deep sleep which lasts for 2 or 3 hours. This stage is followed by light and medium sleep. For some individuals, sound, deep sleep is rare; these light sleepers need more time to get the same amount of sleep than do deep sleepers. It is actually the amount of sleep that counts, but there is at present no unit to measure the sleep amount in the same way that metabolic requirements are measured in terms of calories.

In any case, within the time frame of 24 hours man needs a certain quantity of sleep for energy restoration and revitalization. This must be regarded as a biological law. To break this law by ignoring sleep for 60 to 80 hours leads to extreme fatigue, lapse of attention, irritability, deterioration of performance capability, and finally to visual disturbances and hallucinations. These results have been observed in volunteers in numerous scientific sleep-deprivation experiments carried out in special sleep laboratories. Moreover, dogs kept awake for 14 days showed a degeneration of the ganglion cells of the brain. This should be a serious warning that so-called stay-awake contests, which occasionally have made the headlines in the press, are dangerous and medically not permissible. The same can be said of the use of hallucinatory drugs, which, in addition to their now well-known deleterious psychophysiological effects, deprive the users of 10 to 12 hours for sleep, thus disturbing the biological clock.

History records that both Napoleon I, emperor of France, and Frederick the Great, king of Prussia, tried to prove that sleep is just a bad habit. After two nights, the two rulers had to capitulate to Morpheus, the more powerful god of sleep and dreams. Each experimenter required several days to recover from his loss of sleep, much to the delight of their generals.

Finally, people suffering from insomnia, a prolonged condition of poor sleep or inability to sleep, need medical guidance and treatment. Children especially need adequate sleep regularly because it is mainly during sleep that the pituitary gland produces growth hormones. Long periods of insufficient sleep adversely affect bone and muscle growth.

All this indicates the vital significance of sleep. Aristotle was indeed right when he wrote, 2,250 years ago, "Sleep serves the preservation of the living creatures."

Shakespeare, in *Macbeth*, called sleep poetically and very appropriately:

> "Balm of hurt minds, great nature's second course,
> Chief nourisher in life's feast."

It is, indeed, not surprising that the human body has the tendency to keep its sleep and wakefulness cycle nearly stable, or, more to the point, that the body's clock tends to run at a nearly constant rate.

3

Stability
of the
Clock Mechanism

How stable is this physiological timing device
known as the body clock? Its stability can be ascertained
by observing the behavior of the circadian rhythm under changes of environmental light/dark conditions and
its ability to adapt to shifts in time sequence.

The normal duration of man's circadian rhythm is 24
hours, the time required for the earth to rotate around
its axis. Sleep experiments have proved, however, that
humans are capable of adapting to a slightly shorter or
a slightly longer day. The pioneer researcher in this
respect is Nathaniel Kleitman, professor of physiology
at the University of Chicago, who has the reputation of
having devoted more of his lifetime to the study of sleep
than he has actually spent sleeping. In experiments in the
Mammoth Cave in Kentucky, he found that the physio-

logical clock is able to adapt itself to a light/dark cycle as short as 20 hours or as long as 28. Periods of either shorter or longer duration than those mentioned lead to some discomfort. The biological clock tends to cling to its routine 24-hour rhythm, as is indicated by the behavior of the body's temperature.

Similar experiments have been carried out by other scientists in special sleep laboratories in Germany and France and in the open subpolar regions. One such test took place during the 1960s in the Arctic region near the town of Spitsbergen in northern Norway. A British expedition spent four weeks in this area during a period of constant sunlight. Its members were split into two groups and housed in separate villages, where the groups had no contact with each other or with anyone from the outside except the scientist conducting the experiment.

Each group was given specially designed watches. These appeared to be normal timepieces, but the watches for one group were actually calibrated to run faster than normal, so that in a 20-hour period the passage of 24 hours was indicated. The second group received watches adjusted to run more slowly, so that 28 hours were required to indicate a 24-hour span.

Both groups adjusted to the new time without noticeable difficulty. After eight actual days, the members of the first group believed that they had spent nine days, while the others believed they had spent only seven; neither group was aware of the deception.

A day/night cycle of 20 hours can therefore be considered the medically acceptable minimum and one of 28 hours the maximum. Physiologically the 24-hour cycle is the optimum.

The circadian rhythm of the sleep and wakefulness

pattern persists despite an environment of constant light, as in the instance first cited, or one of constant darkness. This is the natural behavior of inhabitants of subpolar regions, who experience month-long periods of either constant sunlight or constant darkness, and the pattern has also been demonstrated by sleep research conducted in caves and sleep laboratories. The same reactions have been observed in animals and in plants kept under like conditions simulated in the laboratory. Plants that move their leaves up in the morning and down in the evening, and flowers that open and close their petals in the same day/night pattern, continue to do this under artificial constant light or in constant darkness. Forms of animal life whose cycles are tied to the 12-hour period between high tides, such as the fiddler crab, have also been observed to keep this rhythm of activity even when placed in an aquarium.

Shift Work and the Circadian Rhythm

A phase shift of the physiological circadian rhythm is possible, but a certain amount of time is required for readjustment. This is a familiar problem to shift workers in coal mines and factories, as well as in hospitals, transportation, and telephone services. This type of phase shifting is particularly familiar to workers in rocket launching centers, such as Cape Kennedy in Florida and the manned flight-control center in Houston, Texas; when a rocket is being launched, work in shifts around the clock becomes necessary. People involved in these activities requiring other than normal working hours suffer more than inconvenience when they are required to change their work-time schedules.

Conversations with taxicab drivers, for instance, reveal that none of them likes to change the duty time; they prefer to remain either night drivers or daytime drivers. The same is true of policemen, firemen, nurses, and physicians.

In this respect, I should like to mention an extreme example. A night nurse in Frankfurt, Germany, during the second World War, had worked steadily without a vacation for four years, never changing her shift. When, after the war, she was able to take a vacation, she returned to work after the third day. She explained that it was not a vacation at all, because when she was awake and able to enjoy the company of her relatives and friends, they went to sleep. When she became sleepy, on the other hand, she could not tolerate the noise of her surroundings. Apparently, by not changing her work cycle for four years, she had become too deeply adapted to the night duty cycle to adjust readily to a change. Actually she acted too soon; she should have waited several more days.

There are, of course, differences in sensitivity to a phase shift. Some people can sleep like a cat, at any time, at any place, and under any conditions, but most are definitely sensitive in this respect. This is even apparent in the public confusion that results after the 1-hour shift from standard time to daylight saving time and back again.

Rhythmostasis

The circadian rhythm in man and in numerous animals definitely shows a high degree of stability, manifested in keeping its duration and resisting a time shift. This characteristic of the physiological clock can be expressed by the terms *rhythmostasis* or *cyclostasis*, applying the term "stasis" as it was used by Dr. W. B. Cannon in his book *The Wisdom of the Body* (1929). Dr. Cannon used the term "homeostasis" to describe the body's tendency to keep the physical and chemical properties of the intercellular body fluid nearly constant. If homeostasis can be used to describe the control of the internal environment, then rhythmostasis can be used to describe the control of internal rhythms. Certain relations are found between these two features, insofar as some of the circadian rhythm variations in such body functions as respiration, cardiovascular activity, growth, and others fluctuate around their homeostatic baselines. Thus, the tendency to maintain rhythm stability can be regarded as part of the wisdom of the body. In this sense, the term rhythmostasis is definitely applicable to the circadian rhythm, no matter which of the various theories about the cause of sleep and wakefulness is correct. A *rhythmostat* might be said to be located in the interbrain, just as the term thermostat is used to describe the body temperature regulator also found there.

The concept of rhythmostasis is particularly descriptive and useful in the present jet and space age. In air travel, millions of people experience a phase shift of their circadian rhythm by a rapid move from one time zone to another. In space, astronauts are confronted with environments in which there is no familiar time pattern of day and night.

4

Air Travel
and the
Body Clock

IN PRETECHNOLOGICAL TIMES, the power used in travel
was provided by the horse, mule, camel, and elephant.
With the advent of motorized transportation, available
power has increased to thousands and millions of times
the power of one horse. Correspondingly, man's travel
speed has been multiplied manyfold, making the
earth's atmosphere and the near vacuum of space surrounding the earth a medium for travel. The fantastic
progress in the development of travel speed is illustrated
by the title of Jules Verne's book, *Around the World in
Eighty Days* (1878), that of Wiley Post's, *Around the
World in Eight Days* (1931), and now by an expression
appropriate to the space age, "Around the world in eighty-
eight minutes," which refers to the fastest orbital speed
in near-earth space orbit, at an altitude of 200 kilometers.

The effects of space travel on the body clock are considered later. The number of individuals concerned, however, will always be limited. Air travel in the earth's atmosphere is another matter. In 1968 the airlines in the United States alone recorded 150 million people on their lists of travelers. It is difficult to estimate the number of persons who crossed four or more time zones, but a figure in tens of millions would probably not be an exaggeration. This makes the time-zone change by air travel and the effect it may have on the human body a matter that should be of general public interest.

It might be appropriate to begin with an experience of my own. As mentioned in the Foreword, this experience first aroused my interest in the effects of time-zone crossing on the body clock.

In 1949 I traveled by air from Texas to Germany, with a day's stopover in Washington, D.C. After 17 hours of flight via Labrador and Iceland, I arrived in Frankfurt at noon. Following the evening meal, I became sleepy, went to bed and fell asleep immediately, around 8 o'clock. After an hour, I awoke. I was still awake at 11, 12, 1, and 2 o'clock. This seemed very unusual to me, but on thinking it over, I realized that, when I fell asleep at 8 o'clock mid-European time, it was 1 p.m. in Texas, the time of my usual afternoon nap. I finally fell asleep again at about 3 a.m. When I awoke the sun was shining. In the hotel restaurant, I asked the waitress for the breakfast menu. "Breakfast?" she asked. "It is three o'clock in the afternoon. I am sorry; we serve breakfast only in the morning."

But I was pleased that I had stumbled upon an interesting physiological time problem, for 3 p.m. mid-European time corresponded with my breakfast time in Texas. After about a week, I had adapted to the European time. Three weeks later, when I returned by air to the United States, I experienced the process in reverse. I fell asleep early in the evening and awakened around 4 o'clock in the morning. After a week, my sleep time was again in tune with U. S. Central Standard Time. Thereafter, I interviewed numerous transatlantic and transpacific air travelers about their time zone experiences and wrote a medical paper on the subject, which was published in 1952.

At that time, I had no idea that about 20 years earlier the global flier, Wiley Post, together with Harold Gatty, had published *Around the World in Eight Days*. This barnstormer, Post, was actually the first to recognize the significance of the time zones and their effect on the sleep-work patterns and the flying efficiency of pilots. He even suggested ways and means of minimizing the adverse effects of time-zone changes, having experienced these effects firsthand in many long flights in his monoplane Winnie Mae.

Since then, particularly after the introduction of jet propulsion, numerous papers have been published about the physiological effects of flights across time zones. These papers are based on theoretical considerations and on actual inflight and postflight experimental studies carried out by the Aerospace Medical Division of the United States Air Force and the Federal Aviation Administration. Similar studies have been made in Germany, England, Ireland, Holland, France, Spain, Mexico, Japan, India, Russia, and other countries that have air forces, international air lines, or both.

Before discussing the behavior of the body clock in air travel, it will be useful to review the range of present-day travel speeds and the geography of time zones.

Speed Spectrums of Ground and Air Travel

The average long-distance speed of such ground vehicles as automobiles and trains is about 100 kilometers per hour. The average speed of an ocean liner is 30 nautical miles or 55 kilometers per hour. Most jet planes today fly at subsonic speeds—speeds less than the speed of sound, which is 1,400 kilometers per hour. Subsonic speed is also called the first aeronautical speed.

After the sound barrier was broken in October 1947 by Charles Yeager, supersonic speed—the second aero-speed—came into the picture. Aircraft have been flown at speeds of Mach 2 (twice the speed of sound), and even up to Mach 3. These speed ranges may become relatively common for conventional air travel of the future, although they may be restricted to certain regions because of the sonic boom. In juxtaposition with the geographic time zones, the speeds of the jet age create some problems for the physiological clock.

Time Zones and Zones of Seasons

The surface of the earth's nearly spherical globe is subdivided into zones of longitude by 360 lines running from pole to pole. These are called meridians, with the longitudinal line at Greenwich, England, serving as the zero or prime meridian. During the earth's rotation, solar light moves from one meridian to the next in a matter of 4 minutes. The geographic time difference on the earth's surface, then, is 1 hour for

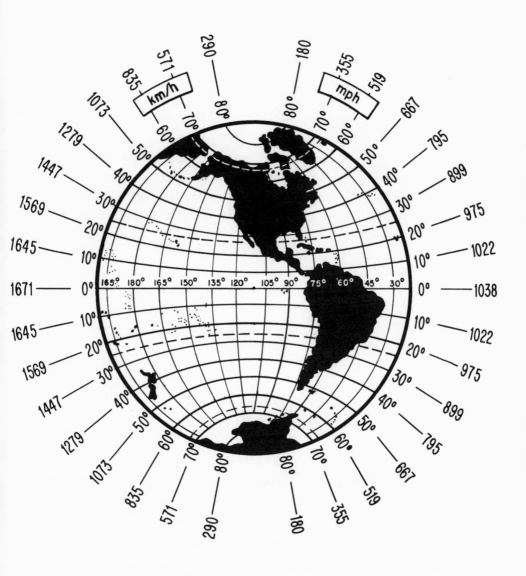

The speed at which daylight proceeds around the world at various latitudes, expressed in kilometers per hour and miles per hour

each 15 meridians. This represents one time zone, and the total number of time zones around the planet is 24. The rapid crossing of a large number of time zones leads to a significant shift in the day/night cycle.

The earth's surface is further subdivided into latitudes by horizontal lines called parallels. The equator is the zero parallel and the parallels measure the distance north and south through 90 degrees. Since the axis of the earth is tilted at an angle of 23.5 degrees in relation to the plane of its orbit, each hemisphere is inclined toward the sun during half the year and away from it during the other half. This causes the seasons, with their variations in the length of days and nights. When it is summer in the northern hemisphere, it is winter in the southern hemisphere, and vice versa. Thus, traveling by air from northern latitudes across the equator to the south leads to a fast change in season. Traversing a large number of both longitudes and latitudes results in a shift of the day/night cycle plus a shift in the cycle of seasons. Of primary importance is the shift of the day/night cycle.

Phase Shift Between Geographic and Physiological Cycles

Within the higher ranges of subsonic speed and in supersonic speed, about half a dozen time zones can be crossed in 6 hours or less. This exposes a traveler in a very short period to a day/night cycle considerably different in time from that at his point of departure and consequently different from the physiological day/night rhythm to which his body is adjusted. This results in a phase shift between these two cycles—the geographic and the physiological. Flight in an easterly direction advances and flight in a westerly direction retards the geographic day/night cycle.

On a westbound flight, with a speed of about 1,000 kilometers per hour—a speed close to the rotational speed of the earth—the sun appears to stay at almost the same place in the sky for the duration of the flight. For a few hours, the traveler experiences a situation somewhat similar to that occurring during daytime on the moon, where a day lasts about 14 earth days. In a flight faster than the rotational speed of the surface below, the sun would be left behind and the traveler would encounter the strange spectacle of a "sunset in the east." During an eastbound flight with high subsonic speed, the day/night cycle is shortened to about one-half. For a few hours this would be similar to the day/night cycle on the outer planets, such as Jupiter and Saturn, which have rotation periods of 11 to 15 hours. However, it is not the solar light condition during the flight but rather the situation after landing at a different time zone that causes problems for the traveler.

The extent of the time-zone shift becomes obvious if one looks at the meridians on a terrestrial globe and then at the routes and flight schedules of the national and international air services.

Flights from coast to coast in the United States cross four time zones. Transatlantic flights, which began with the historic flight of Charles A. Lindbergh in 1927, cover five or six time zones. A flight from California via the subarctic to the Scandinavian countries includes eleven time zones, and transpacific flights from the west coast of the United States to Japan cross eight. Crossing twelve time zones, as one does in traveling from New York to New Delhi, leads to a complete

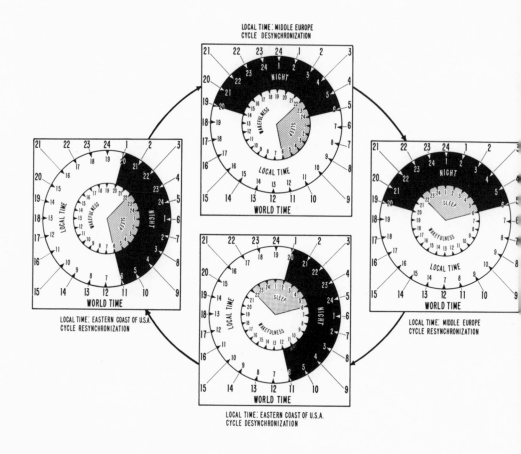

Phase shift of the circadian cycle
  (above) flight from the eastern United States to Europe and back
  (right) flight from the eastern United States to Japan or Australia and back

54

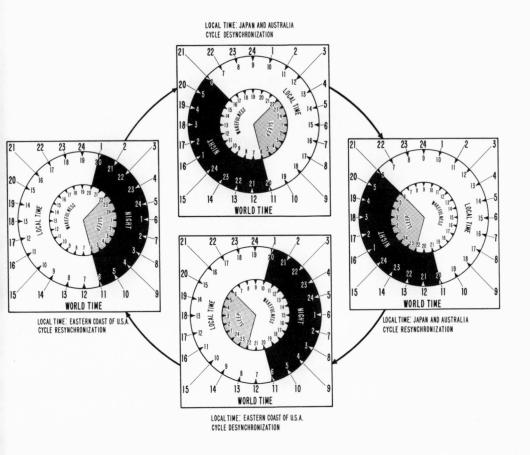

The outer area of the squares shows Greenwich Time (universal time). The outer large circle gives the local geographic time and the smaller inner circle the physiological time. The black section of the outer circle indicates night and the shaded section of the inner circle the normal time of sleep.

cycle inversion. When it is noon in India it is midnight on the eastern coast of the United States.

Obviously, long-distance flights along, or nearly along, the latitudes—westbound or eastbound—cause a noticeable shift if four or more time zones are crossed. The traveler's body clock, on arrival, will still be in phase with the time of the place of departure and will not coincide with the astronomically determined local time. This unnatural relationship between the internal time of the traveler and the external local geographic time is called desynchronization, or desynchrony. It results in a phase shift between the geographic cycle and the physiological one. As mentioned earlier, flights in an easterly direction advance, and in westerly direction delay, the day/night cycle. Several days to a week may be required for the physiological cycle to become adapted to the local time at the termination of the trip. After adaptation, the two cycles are described as resynchronized.

This effect is graphically demonstrated in the diagrams on pages 54 and 55.

The first set of diagrams presents the situation before and after a flight from the east coast of the United States to Central Europe and back. The diagram at left indicates the physical and physiological day/night cycle in New York or in Washington, D.C. The diagram at top center shows the geographic time change by 6 hours on arrival in central Europe. The physiological circadian cycle is still the same. Three to 6 days later, it is resynchronized, and is again in tune with the local physical day/night cycle, as shown in the diagram at right. Reading from the diagram at right through the one at bottom center to the one left demonstrates the behavior of the physical and physiological cycles after a trip from central Europe to the east coast of the United States.

The second set of diagrams shows the corresponding geographic and physiological time pattern of a trip from the east coast of the United States to Japan or Australia and back. These two points have a time difference of about 10 hours.

Statistical studies after long-distance flights have shown that the majority of people, older people especially, are sensitive to this travel-produced phase shift and experience some discomfort. They become hungry, sleepy, or wide awake at the wrong time in relation to the new local time. Their head clocks, their stomach clocks, and their elimination systems are confused. This is, in brief, the picture of the circadian phase-shift syndrome.

After transcontinental flights in the United States, this condition lasts from 3 to 4 days; after transatlantic flights, it lasts 5 or 6 days. After crossing twelve time zones, which constitutes a complete reversal of the day/night cycle, resynchronization may take as long as 10 to 12 days. These figures demonstrate clearly the resistance of the body clock to a sudden major time change. Dr. William Douglas, flight surgeon to the Project Mercury astronauts, and an Air Force medical officer with rich experience in long-distance flights, has suggested an easy-to-remember general rule:

Most travelers adjust to a new circadian cycle at the rate of one hour per day.

Any rule of course presupposes variations. Some people feel that they become adjusted more easily after eastbound flights; others, after westbound flights. Some feel that they adjust more easily after returning to their home time zone with its familiar geographic climate

57

and social milieu. Solitary travelers may take longer to adjust to the time difference than do those traveling in groups. At present, there is not enough statistical material available to give scientific reasons for these variations. There are some people who are not particularly sensitive to time-zone changes, just as there are some who adjust relatively quickly to shifts in work time. Babies during their first three months, when they are in the minicyclic stage, show no effects; however, after they have become monocyclic and light-active, they should show some of the symptoms of desynchronization when transported across several time zones.

If a term related to medical language is needed to designate the psychophysiological effect of cycle desynchronization, *desynchronosis* would be an appropriate one, and the individual could be described as *desynchronotic*. This condition, however, is not an illness or a pathological condition, but merely a time disharmony between what the body's internal physiological milieu expects from the external physical and social milieu at the new locale, and vice versa. Expressed more musically in clock language, it is a disharmony between the tick-tock of the physiological clock and the ding-dong of the city's tower clock. This disharmony of time can be significant in many respects.

Significance of Cycle Desynchronization

Circadian cycle desynchronization can have some significance in relation to international political conferences. During the first few days of such a meeting, the individual who has crossed a number of time zones may be somewhat handicapped by temporary lapses in alertness resulting from his desynchronotic condition. This certainly should be of interest to heads of state and their advisers called upon to participate in summit meetings. Durng emergency meetings of the United Nations during the past 26 years, the representatives of one side remained in their own zone, or close to it, while those on the other side had to cross five to ten time zones. Historians might find it interesting to take a "physiological clock" look at these meetings and at some of the participants' strange and extravagant behavior, such as pounding on a table with a shoe.

Executives of large companies who frequently visit their subdivisions abroad, and scientists attending international congresses at places on the other side of the oceans, are faced with the same situation, as are athletes participating in Olympic games, concert musicians, and stage or film actors, all of whom frequently have to crisscross a multitude of meridians.

The problem of circadian cycle desynchrony is especially important for those whose occupation constantly involves time-zone changes. Pilots and other personnel on long-distance air routes are in this category. They have to cross and recross a number of time zones several times a month. They may even have to fly around the world once a month. A too frequent shift of their physiological circadian rhythm causes nervous strain and fatigue and requires special attention. This is well understood by pilot associations, airline medical directors, and the medical officers of the air forces, and measures are taken to reduce the effects.

So far, this discussion of the circadian rhythm and time zones has dealt only with the effects on persons in a healthy state. When a traveler is ill, there are additional aspects. For instance, a traveler with fever, on arriving from a location some five or more time zones away, has the low and high of his body temperature in the temporal pattern of the preflight time zone. Arriving in New York from Europe, a traveler who has a cold with fever may have the maximum temperature not at 5 o'clock in the afternoon Eastern Standard Time, but at 11 o'clock in the morning, which is 5 p.m. mid-European time. This factor is diagnostically important. There are other clinically interesting points concerning medical therapy, or the best time for surgery, on a patient who has just arrived from a distant zone, but these are of concern only to medical men.

Minimizing Time-Zone Effects

What can be done to prevent undesirable effects of a phase shift in the circadian rhythm after long-distance flights? There are several ways to avoid desynchrony or to shorten the period required to become synchronized with the local time of the destination.

First, if you plan to attend an important meeting at a distant location, you can preset your physiological clock by adopting, several days before departure, a sleep/wakefulness time pattern corresponding to the day/night cycle of your destination. This is preflight adaptation or synchronization. If you expect to travel in an easterly direction, you can accomplish this by going to bed an hour earlier each evening, progressively, and arranging to be awakened an hour earlier each morning until you have moved your own sleep schedule forward to coincide with the local time at your point of arrival. If your trip is to be westbound, you should reverse the procedure, retiring an hour later and arising an hour later each day.

Second, you can fly to the distant place several days in advance of the event or meeting, giving the body time to adjust after arrival. This is postflight local adaptation or synchronization. President Dwight D. Eisenhower did this when he had a summit meeting with Nikita Khrushchev in Geneva on a Monday in 1955, he arrived on the preceding Friday.

Either the preflight or postflight method should be effective in keeping the traveler awake and alert the entire day of his engagement.

The traveler who cannot take the time for adaptation must keep in mind that the morning hours during the first few days after long-distance eastbound flights, and the late afternoon hours after westbound flights, are not the proper times for important negotiations and decisions and conduct his affairs accordingly. Some industrial companies in the United States forbid their executives to sign contracts the first two days after transatlantic or transpacific flights.

Finally, a mild medication taken at the proper time and in proper dosage might be helpful, as a kind of biochemical synchronizer, in physiological adjustment to the new local geographic day/night cycle. The application of electrically induced sleep has also been suggested for the same purpose, but more knowledge of the physiology of such sleep and the development of instrumentation for producing it are still needed. Natural means, such as physical exercise and warm baths, are certainly preferable to artificial means to shorten the waiting time for sleep.

While the problem of desynchronization is less important for vacationing travelers, they too suffer the effects in greater or less degree and should make allowance for these in their planning. Postponing a strenuous sight-seeing program until the body clock is at least partly synchronized with local time will make for a more satisfactory holiday.

It is conceivable that an occasional new setting of the body's clock, as is the case after long-distance flight, might have some stimulating effect on the clock's system and serve as a kind of cycle exercise. Although there is not yet any proof for this theory, the possibility should not be ignored.

The measures described for cycle coordination are useful and applicable if the traveler stays at his destination for several days. The higher jet speeds, however, can also facilitate coordination in so far as they permit round-trip flights to distant points within the same day —for instance, flights from one coast of the United States to the other and back again. In this case, phase shift occurs within a 24-hour period and should cause no desynchronization problems. This is especially true at supersonic speeds, which would make possible a round trip from Washington, D.C., to the capital cities of Europe within a day.

Time Zones and Sea Travel

So far, the physiological clock has been considered only with regard to high speeds provided by aircraft. The comparatively low speeds of surface vehicles can also lead to some interesting and pertinent experiences. Modern ocean liners, having a speed of about 30 nautical miles or 55 kilometers per hour, can cross the five time zones of the Atlantic in 4 to 7 days. Consequently, the ship's time is changed by 1 hour every day, or every 1.5 days. This daily change has a noticeable effect on the behavior of the passengers, according to observations I have made during half a dozen transatlantic ship crossings. After passing the mid-Atlantic when traveling from the United States to Europe, more and more passengers are late for breakfast. On large ocean liners, with many passengers, there are usually two sittings for meals. The first breakfast may be at 7 o'clock and the second at 8. On an eastbound Atlantic trip, it is advisable to choose the second sitting for breakfast; on the return trip, the earlier one.

One woman told me of her experience in returning from Germany to the United States with her two-year-old daughter. Because she preferred late breakfasts, the mother chose the second sitting. However, she soon found that by lunchtime each day the toddler was ready for her afternoon nap and consequently went to sleep in her high chair, her lunch untouched. Daily, the mother smuggled rolls and fruit to their cabin so the child could have a snack in the afternoon, resolving that next time she would choose the early sitting.

After midway on an eastbound trip, many passengers stay up very late at night, according to the ship's time, crowding the decks and bars. Thus, after landing, their body clocks are well prepared for the night life in Paris, Munich, and London. On the return trip from Europe, passengers tend to go to bed earlier each night. There are usually not too many late-evening social activities on the ship during the second half of the journey. On trips from the west coast of the United States to the Far East and back, this picture is reversed.

Migrating Birds and Time Zones

The flying speed of birds is in the range of 30 to 150 kilometers per hour, or about the same as that of motorized surface vehicles. It is interesting to note that most long-distance migratory birds do not cross more than two or three time zones on their seasonal north-south and south-north flights. Ornithologists have mapped four major migration flyways on the American continent: the Eastern, Mississippi, Central and the Western flyways. These more or less overlap from two to four of the eight geographic time zones of the two American hemispheres. Many migrators have their nesting places in Alaska and northern Canada during the summer, and their winter residence in the southern regions of the United States, in Mexico, and in the northern regions of South America. Prominent representatives of this group are the swallows of Capistrano in southern California (Western Flyway) and the whooping cranes in Port Aransas, Texas, on the Gulf of Mexico (Central Flyway). Many migrants use the Mississippi Valley. The golden plover travels from northern Canada along the Eastern Flyway to Brazil and Argentina and back. Through this habit of limiting their flight paths to a minimum of time zones, birds avoid drastic changes in the length of daylight periods.

The European migrating birds—storks and cranes—stay on their flyways from northern and central Europe to southern France, Italy, Spain, and northern Africa within the range of one to three time zones. Three time zones are crossed by those birds which avoid the high Alps and the Pyrenees or the Mediterranean Sea by flying around them. Birds of the northern Russian forests fly to the Black Sea and Turkey, crossing only two time zones.

There are, of course, exceptions. The flight route of the Arctic tern, which summers in the Arctic and winters in the Antarctic, goes via Labrador, Iceland, Scotland, the Atlantic coast of France, and the West African coast to the Antarctic. It travels a total of about 18,000 kilometers. This long-distance champion migrator crosses five time zones, but it stops frequently for rest on islands and coastlines, thus minimizing the effects of time-zone changes by gradual adaptation. Another exception is the Norwegian snow goose, which flies from Norway to Labrador and back. This bird is not affected by time-zone changes because its route remains in the twilight zone of the Arctic.

That birds and mammals are sensitive to a phase shift in their sleep-and-activity cycle is a familiar observation in zoological gardens with animals imported from distant countries. For a week or so, the animals are sleepy and do not care to look at the zoo's visitors. Gradually, they become alert during the visiting hours, exhibiting interest in the strange, upright-walking creatures who keep them behind bars and under artificial light until late in the evening, a condition unknown to them in the wilderness of Africa, Australia, or New Zealand.

I would like to add an example with which I was personally concerned. One morning, some 15 years ago, I heard in the news report of a San Antonio radio station that several pelicans had died in England's Buckingham Palace gardens and had been replaced by three pelicans flown from Corpus Christi, on the Gulf of Mexico, to the London Zoo. But the head keeper of the zoo was not satisfied with their behavior. He was quoted as saying, "I have never seen such sleepyheads as these Texas pelicans." I immediately wrote to the head keeper, explaining that this behavior could be expected, since the birds, after crossing seven time zones, still had the Texas time in their systems, and it might take several weeks for their biological clocks to run in tune with Big Ben, the clock on the Tower of the Houses of Parliament. Three weeks later I received a friendly reply informing me that the pelicans were doing well, that they were awake and alert during the daytime, and well prepared to join the other animals in the royal gardens. The head keeper was pleased, and so was I, that the honor of the Texas pelicans was restored.

Bird migration probably started with the disappearance of the last ice age, some 70 million years ago, and has evolved into an instinct. This miraculous phenomenon is governed by outside environmental light and temperature factors and the availability of food. It is also prompted by impulses inside the birds' bodies, such as mating instincts. These are controlled, according to ornithological theories, by hormones primarily of the pituitary gland. This control is also indicated by the fact that migratory birds accumulate fat in their tissues several weeks before their departure. This fat provides them with biological fuel, so to speak, for their long air journeys. But, as mentioned, they keep their flyways within distinct longitudinal limits.

5

Adjustment
to
Space Flight

Subsonic and supersonic speeds have already been discussed in relation to the global time zones and the resulting physiological clock implications. These two aeronautical speeds, made possible by propeller-driven planes and jet planes, were extended in the late 1950s to astronautical velocities produced by rocket propulsion.

The first astronautical or cosmic velocity, on the order of 8 kilometers per second, permits orbital or satellite flight in near-earth space. In this new situation the customary geographic day/night cycle is replaced by a short, erratic sunlight/shadow cycle.

The Light/Dark Cycle

Within the relatively radiation-safe altitude range of from 200 to 800 kilometers, which is below the Van Allen radiation belt, the orbital flight periods last from 90 to 130 minutes. About 40 percent of this time, depending on the orbit's inclination, is satellite light, or earth-shadow time, as it can be called. This external light/shadow or photoscotic cycle in orbital space flight is not longer than one-tenth of the day/night cycle on earth. Furthermore, it is modified by earthshine and moonshine, with a permanently velvet-black sky in the background.

But in this short periodic, exotic light environment encountered during flight in near-earth space, the astronaut still is bound to the temporal pattern of sleep, rest, and activity of his inborn circadian rhythm. The behavior of his physiological clock continues to be dictated by his physiological nature as a terrestrial creature whose ancestry goes back for millions of generations. Before manned space flight began, doubts were occasionally voiced about the possibility of sleep under the condition of weightlessness. Now a better picture of the actual situation with respect to weightless sleep has been obtained by evaluating the sleep experiences of some of the astronauts who have been in near-earth space for 24 hours or longer. The telemetric recordings of the "medical space practitioners" in charge of the medical control of the flights have also provided considerable data.

In the earlier manned space flights, sleep was recorded by means of electroencephalograms, which require electrodes to be placed on the head of the astronaut. The electrodes were found to be uncomfortable, and this method has been abandoned. The recorded heart rate, respiration, and blood pressure now serve as the indicators of sleep to the observers on earth.

Experiences of Astronauts and Cosmonauts

During the last mission of Project Mercury in 1963, which consisted of twenty-two earth orbits and lasted 34 hours and 20 minutes, astronaut Gordon Cooper found, even early in flight, that when he had no tasks to perform and the spacecraft was oriented so that the earth was not in view from the window, he easily dozed off for brief naps. During the period scheduled for sleep, he slept only in a series of naps lasting no more than about an hour each. His total sleep time was about 4.5 hours. He told me that he could have slept for much longer periods if there had been another man on board to monitor the systems. He also stated that he slept perhaps a little more soundly than he did on earth.

The first "cosmic slumber" of the second Russian cosmonaut, Gherman Titov, who in 1961 experienced seventeen orbits in 25 hours and 18 minutes, was not without interruptions. After seven orbits, he felt that he was in a definite state of fatigue. When he flew over Moscow at 6:15 p.m., he prepared for sleep in accordance with the schedule. He released special belts from the side of the seat, strapped his body to the contour seat, and adjusted the seat to the bed position. He promptly fell asleep but awoke, much earlier than scheduled, during the eighth orbit. When he opened his eyes, he saw his arms dangling weightlessly, and his hands floating in midair.

"The sight was incredible," Titov reported in his

book, *I Am Eagle* (1962). "I pulled my arms down and folded them across my chest. Everything was fine until I relaxed. My arms floated away from me again as quickly as the conscious pressure of my muscles relaxed and I passed into sleep. Two or three attempts at sleep in this manner proved fruitless. Finally, I tucked my arms beneath my belt. In seconds I was again sound asleep." According to Titov: "Once you have your arms and legs arranged properly, space sleep is fine. There is no need to turn over from time to time as a man normally does in his own bed. Because of the condition of weightlessness, there is no pressure on the body; nothing goes numb. It is marvelous; the body is astoundingly light and buoyant . . . I slept like a baby."

He awoke at 2:37 a.m. Moscow time and found himself a full 30 minutes behind schedule because he had overslept. He immediately started the required morning calisthenics. Thereafter, he carried out all scheduled assignments with no problems, and during the seventeenth orbit prepared the rocket and himself for the "baptism of fire," the atmospheric re-entry.

Titov's sleeping period coincided largely with night time over Russia, as did those of the other Russian cosmonauts.

In 1963, Valery Bykovsky, during his 119-hour flight of eighty-one orbits, slept four times for periods of 8 hours, alternating with periods of 16 hours of wakefulness. Valentina Tereshkova, during her 71-hour flight of forty-seven orbits in 1963, had much the same sleep experience. According to Russian scientists, during both of these flights, the diurnal periodicity of physiological functions changed only during the first and last days of the weightless state; this was probably related to the emotional strain. During the phases of wakefulness, brief rest periods were usually scheduled for times when the spaceship was not over the Soviet Union. The scientists also noted that at night, during sleep, nearly all cosmonauts showed a greater reduction in pulse rate than that recorded during the same hours in previous simulated flights. This reduction in pulse rate is an indication of good and sound sleep.

The three-man team of the spaceship Voskhod, in 1964, rested and slept in shifts during their 24-hour, 17-minute flight.

In 1965, there were two orbital flights in the Gemini series by American astronauts, in which special attention was given to the sleep-and-wakefulness cycle. Neither the Gemini 4 crew (James A. McDivitt and Edward H. White) nor the Gemini 5 crew (Gordon Cooper and Charles Conrad) reported any difficulty in performance resulting from the 45-minute darkness-and-daylight cycle created by orbital flight, according to Dr. Charles A. Berry, Chief of Medical Space Operations, NASA Manned Space Center. "The GT-5 crew had a long sleep period, programmed in conjunction with night time at Cape Kennedy. They, too, had intermittent spacecraft noise irritants interfering with sleep, and found themselves tending to retain their ground-based, day/night cycle." It appears best, Dr. Berry notes, "to provide a joint long-sleep period related to normal sleep time at the 'Cape.'"

During the 14-day orbital flight of Gemini 7, which was the longest to date by American astronauts, Frank Borman and James A. Lovell had no significant sleep difficulties. The inside of their spacecraft was artificially darkened by coverings on the windows, providing a microenvironmental day and night of their own, which was kept in tune with the day/night cycle at Cape Kennedy.

The program of the group flights of the Russian Soyuz 6, Soyuz 7, and Soyuz 8 in October 1969 included 8 hours of sleep in each 24-hour period; however, the cosmonauts actually slept only 6 or 7 hours and subjectively regarded this as fully adequate because they felt refreshed and fit for work. The sleep time was kept more or less in tune with the night time at the launching area near the Ural Mountains. In contrast, the two cosmonauts in Soyuz 9 during their 17-day, 17-hour record flight in January 1970, had to shift their sleeping cycle by 12 hours because their spaceship passed over Russia at night and they had to land in the early morning. They slept during the daytime and worked at night. "Our sleep in flight was normal. After the sleep we felt refreshed and quite efficient," cosmonaut Andrian Nikolayev, the commander of Soyuz 9, told me.

By and large, the recorded and reported sleep-and-wakefulness time patterns in orbital space flight reflect the physiological circadian rhythm of 24 hours. But for the astronauts, the time zones on the earth's surface are environmentally meaningless, since they cross one time zone every few minutes. Their basic guiding time is Greenwich Time or Universal Time, but their body clocks have to be coordinated more or less with their natural circadian rhythm. For flight operational reasons, it is very desirable that these body clocks also remain synchronized with the local time of the Mission Control Center on earth to which they were adapted during the pre-launch period.

Sleep Schedules for Space Stations

During the 1970s, space stations will probably be established for meteorological, geological, astronomical, and biomedical studies. Such a manned orbiting research laboratory, or "sky lab," is being planned by NASA. It will be manned by a crew of three men, who will be ferried to the space station and back to earth at intervals of from 4 to 8 weeks, by means of a reusable space shuttle. This sky lab will be the predecessor of a larger permanent space station, which will accommodate a staff of twelve, including crew members, scientists, and experimenters. The duty and sleep regime of the professional personnel in charge of the flight operation will require a rotating shift, which will be made possible by more comfortable sleeping facilities than are available in space capsules. Switching this shift within the operational team should be avoided, however, since this would cause additional stress at a time when the men must perform demanding jobs. The research passengers will also sleep in special soundproof sleep compartments, in order to remain fresh and alert during the hours when they are engaging in exploratory tasks.

The clock time in the space station will be about the same as that in the Control Center on earth. Time regulation will undoubtedly be a decisive factor for successful research work in these space-bound scientific "institutes."

Sleep and activity problems similar to those connected with the sky lab or space stations were encountered in the Sea Lab and the underwater stations called Tektites. This similarity was illustrated when astro-

naut Scott Carpenter of Project Mercury participated
as an aquanaut in the exploration of the more or less
lightless "inner space" of the deep sea in Sea Lab II on
the coast of California in 1965.

6

Problems of Lunar and Interplanetary Missions

THE FIRST SUCCESSFUL PHASE of manned space flight—orbital operations in near-earth space—was followed by the flight to the nearest celestial body, the moon. In 1969, Apollo 11, after orbiting the earth several times, took off from the gravitational field of earth at an escape velocity of 11.1 kilometers per second, heading for the moon. The light environment en route to the moon includes permanent sunshine, earthshine, moonshine, and a velvet-black sky. After departure from earth, about 3 days are required to achieve a circumlunar parking orbit. The relatively short duration of the transmoon trajectory requires a carefully planned sleep-and-duty regime for the three-man team of lunar astronauts.

75

Findings from Project Apollo

The sleep/work cycles of the crew members of some of the Apollo missions have been described in detail by Dr. Charles Berry. Sleep was somewhat irregular in the first flights but showed gradual improvement. Frank Borman, commander of Apollo 8, which made the first flight around the moon, did not sleep well during the entire flight. One reason, according to him, was that the crew members slept at different times. The noise caused by the activities of the other crew members going about their tasks disturbed him. On the other hand, the astronauts of Apollo 9 (James A. McDivitt, David R. Scott, and Russell L. Schweickart) and of Apollo 10 (Thomas P. Stafford, John W. Young, and Eugene A. Cernan), who slept simultaneously, experienced normal sleep, according to Stafford.

During the Apollo 11 mission, astronauts Neil A. Armstrong, Edwin E. Aldrin, and Michael Collins slept simultaneously on the way to the moon, consequently having good sound sleep. I had the privilege of observing this mission and the earlier ones from the Mission Control Center in Houston. While Apollo 11 was on the moon, Armstrong did not sleep at all, and Aldrin slept only 2 hours. During the Apollo 12 mission, the sleep phase of astronauts Charles Conrad, Jr., Alan L. Bean, and Richard F. Gordon, Jr., had to be postponed about 6 hours after the craft entered the transmoon trajectory for reasons of "orbital mechanics." However, the 3-day duration of the flight gave them time enough to adapt to this shift of the circadian rhythm, so that they were in good condition for the entrance into a circumlunar orbit and for the landing maneuver. They stayed on the moon long enough to have a 7-hour sleep period before lift-off.

During Apollo 13's emergency phase, which lasted 86 hours one of the disturbing factors for astronauts James A. Lovell, Fred Haise, and John L. Swigert was loss of sleep. The sleep or rest cycle of the team of Apollo 14 (Alan B. Shepard, Edgar D. Mitchell, and Stuart A. Roosa) was somewhat irregular because of orbital mechanics; nevertheless, they slept well. During their 34-hour stay on the moon, Shepard and Mitchell had two periods for sleep after their two excursions on the lunar surface. The necessity of integrating the astronauts' sleep pattern into the fixed flight plan has been a rhythmological problem of all the lunar landing missions. Interestingly enough, all the astronauts slept better on the way back to earth than they did on the way to the moon.

Lunar Stations

In a future lunar station or research laboratory, the sleep-and-activity cycle will be completely independent of the physical light/dark cycle of the moon, which is equivalent in length to 27 terrestrial days. During the light part of the cycle, the sunshine has an illuminance, or illumination, of 140,000 lux, which is about the same as that above the earth's atmosphere. (Intensity of illumination is now measured in lux, or meter candles; formerly the unit was the footcandle.) To add to the brightness, however, there is also earthshine, or the reflection of the sun off the earth (the lunar counterpart of moonshine on earth). This earthshine is seventy-five times stronger at "full earth" than is moonshine on earth at full moon. One main difference in the light on the surface of the moon is the absence of an atmosphere, which on earth causes light to be refracted and spread more evenly. The light and dark on the moon's surface are harsher and more pronounced.

Such is the general light picture on the moon. But there are also locations, such as in craters, where no direct sunshine strikes, and other places with constant sunlight, such as on the "mountains of eternal light" located near the lunar south pole.

In a broader sense, the sunrise/sunset cycle on the moon does not provide man with a time cue comparable to the 24-hour dark/light cycle on earth. The moon has about 2 terrestrial weeks of sunshine, followed by darkness of the same duration. Therefore, the astronauts inside a lunar station would have to schedule their sleep-and-activity cycles in terms of their terrestrial circadian pattern. The cycles must be arranged in shifts among the members of the operational team, to keep in constant radio contact with the moon base control center on earth. Communication is possible only when the control center is on the near side of the rotating earth as seen from the moon. Unlike the relatively small Apollo command module, any lunar laboratory will be able to provide separate areas for sleep and work, making it possible for crew members to sleep in shifts without being disturbed.

Generally, sleep on the moon might be better than on earth due to the moon's lower gravitational force, which is only one-sixth of what we are accustomed to on earth. A recording of a sleeping astronaut, made by a hypnograph, an instrument that measures body activation during sleep, will probably show fewer of the body movements which sometimes interrupt sleep. This is because pressure between the body and the bed will be greatly reduced. In contrast, the Russian moon robot Lunakhod 1 had to go into a state of hibernation during the lunar night, which is 14 earth days long. It was functionally alive again for a comparable period for experiments controlled by engineers on earth.

Manned Flights to Mars

The first postlunar planetary target for a manned mission will be the planet Mars, as is envisioned for the mid-1980s by Wernher von Braun, Deputy Director of NASA. A flight to Mars requires complete escape velocity from the gravitational field of the earth—11.2 kilometers per second, or the second astronautical velocity. If this interplanetary journey were based on a minimum energy trajectory, more than 8 months would be required to reach the gravitational field of Mars. From a medical point of view, this period in interplanetary space is too long; it must be shortened to 20 percent of that time before such a voyage becomes possible. New kinds of power, such as nuclear propulsion, are expected to achieve this.

When the spaceship, with a crew of six to eight Mars-bound astronauts, is 1,300 million kilometers beyond the shadow of the earth, it will be in constant sunshine against a velvet-black sky. This means, in effect, constant day and constant night, with the sun as a symbol of the day and the dark sky a symbol of the night. In this exotic, nonperiodic light environment along the trans-Mars trajectory, the occupants of the spaceship must arrange their sleep, rest, and activity regime to correspond with the temporal pattern of their circadian rhythm on earth. This pattern might be what is called

79

in the science of biorhythmology a "free-running cycle," which certainly will include a sleep period of 6 to 8 hours out of every 24. Six hours of sleep with occasional catnaps during each 24-hour period has been found to be adequate in space-cabin simulator experiments lasting from several weeks to about two months. As to the length of the whole circadian rhythm, experiments previously described indicate that it can be shortened to 20 hours or prolonged to 28. The body temperature curve, for instance, adjusts itself to a cycle change of as much as 4 hours either way.

It is encouraging to learn from the experiments already made that space sleep poses no difficulties if conditions within the cabin are adequate. Sleep is a prerequisite for the maintenance of health, and consequently of high performance capability, for the astronauts. Furthermore, exercise, which must be carried on to prevent certain types of physiological deterioration, will automatically contribute to the establishment of a satisfactory pattern of sleep and wakefulness.

Within half a million kilometers of the Martian surface, the Mars ship will be in Mars's gravitational field and can enter a circum-Martian orbit in preparation for the landing maneuver. If an altitude of 100 kilometers is chosen, the occupants will observe a cycle of sunshine and Mars shadow of about the same duration as in the departure orbit in near-earth space.

On Mars itself, the day/night cycle is only 37 minutes longer than that on earth. The daytime sky is dark blue in color, except for an occasional region covered with thin white clouds. Solar illuminance on the Martian surface at noon may reach one-third of that on earth. Thus, the dark/light alternations on Mars offer a time-cue sequence that should be familiar enough to terrestrial visitors for their sleep/activity cycle to be maintained. Since Mars's gravity is less than one-half of that on Earth, there should be no sleep difficulties in a Martian station.

If there should be indigenous life in the form of vegetation in the dark, blue-green surface regions of Mars, thriving on some type of photosynthesis, it would be active only during about 5 daylight hours. At night it would pass into a dormant state, due to the extremely low temperature of the Martian night.

According to modern ground-based and space-bound astronomy, Mars will probably be the only postlunar astronautical target for a manned landing mission. All the other planets present extremely hostile environments. They are either too hot, as are Venus and Jupiter, or too cold, as are the other outer planets. Venus's rotation takes 293 terrestrial days in retrograde direction —that is, opposite to the direction of its revolution around the sun. The slowly rotating planet Mercury, closest to the sun, combines both temperature extremes. It is too hot on the sunlight side and too cold on the opposite side.

Some billion kilometers beyond Pluto in interstellar space, solar illuminance drops below the light minimum for reading and color vision. This is a realm of eternal night, with the sun attaining a stellar magnitude not much different from that of other stars. In interstellar flight, which requires the third astronautical velocity (the escape velocity from the gravitational field of the sun, in which a velocity in higher fractions of the speed of light is required), the sleep and wakefulness cycling of interstellar space travelers must be imagined as being projected against the phenomenon of time dilation or prolongation. Their physiological

cycle might be many times longer than the corresponding cycle on earth, but they would not be aware of this.
The reason is that, at speeds approaching that of light,
molecular movement slows down. This factor would
affect the body functions, including the biological
clocks, of interstellar travelers. But while interstellar
flight is a matter of fantasy and science fiction,
manned interplanetry flight as far as Mars including
the necessary adjustments of man's body clock, is in the
realm of reasonable, realistic science vision.

In conclusion, the biological clock with its circadian
sleep-and-wakefulness rhythm is playing an increasingly
important role in man's life in his home milieu on
earth, as well as in air travel across time zones and in
the conquest of space. Furthermore, the study of the
mechanism under conditions of weightlessness in space
may shed new light on the true nature of this mysteri-
ous timing device. Any such findings would be a bene-
fit to man on his native planet.

Conversion Table

Glossary

Bibliography

Index

Conversion Table

LENGTH

1 mile (mi), statute or land = 1.609 kilometers
1 nautical mile = 1,852 kilometers
1 kilometer (km) = 0.621 statute mile
1 kilometer = 0.540 nautical mile

TEMPERATURE

To convert Centigrade to Fahrenheit, multiply by 9, divide by 5, and add 32 degrees.

To convert Fahrenheit to Centigrade, subtract 32 degrees, multiply by 5, and divide by 9.

Glossary

biorhythmology	The science of rhythms in living beings
biosphere	The living world
chronobiology	Another term for biorhythmology
circadian	About one day (Latin *circa*—about or around, *dies*—day)
cyclostasis	The tendency of the body to keep its cycles nearly constant
diurnal animal	Daytime active animal
electrocardiograph	Device which records the electric currents produced by the heart muscle (electrocardiogram: EKG)

electroencephalograph	Device which records the electrical currents of the brain (electroencephalogram: EEG)
endocrine glands	Glands producing one or more internal secretions or hormones
endogen	Developed internally
exogen	Produced by external condition
exteroceptor; exteroreceptor	Nerve ending that responds to external environmental stimuli
homeostasis	The tendency of the body to keep the physics and chemistry of its fluids nearly constant
hormones	Chemical substances produced in endocrine glands which have specific effects
hypnograph	Device attached to the mattress of a bed to record the body movements during sleep
illuminance	The intensity of light coming from a light source, measured in meter candle or lux
interoceptor; interoreceptor	Nerve ending that responds to stimuli from the internal organs
intradian	Within a day
isochronous	Taking place at equal intervals of time
lux	Unit for measuring the intensity of illumination; 1 lux = 1 meter candle
macrocosmos	The universe at large
microcosmos	Space of microscopic size
microsleep	Sleep seizure lasting only a few seconds
nocturnal animals	Nighttime active animals

Glossary

polycyclic	Having many cycles
polyrhythmic	Having many rhythms
retrograde movement	Backward movement
revolution	Movement of a celestial body around another celestial body
rhythmostasis	The tendency of the body to keep its rhythms nearly constant
rotation	Movement of a celestial body around its axis
sonic boom	A noise caused by a shock wave or pressure disturbance that emanates from an aircraft at and above the speed of sound
speed of light	299,792 kilometers or 185,971 miles per second
speed of sound	1,400 kilometers or 740 miles per hour
subsonic speed	Speed below the speed of sound
supersonic speed	Speed higher than the speed of sound
synchronize	To cause to operate at the same time or rate
synchronous	Operating or taking place at the same time
time dilation	Prolongation of time. According to Einstein's theory of relativity, time would slow down for occupants of a spaceship moving at velocities approaching the speed of light. This would not be noticed by them until they return to earth.
universal time	Greenwich time
Zeitgeber	Time cue for entraining sleep and wakefulness

87

Bibliography

Akert, K.; Bally, C.; and Schade, J. P., eds. *Sleep Mecha-nisms. Progress in Brain Research*. Amsterdam, London, and New York: Elsevier Publishing Co., 1965.

Aschoff, J. "Zeitgeber der tierischen Tagesperiodik," *Naturwissenschaften* 41:49-56, 1954.

Bedwell, T. C., and Strughold, Hubertus, eds. *Third International Symposium on Bioastronautics and the Exploration of Space*. Brooks Air Force Base, Texas: Aerospace Medical Division, 1964. See especially J. Aschoff, "Significance of Circadian Rhythms in Space Flight"; O. G. Gazenko, "Medical Investigations of Spaceships Vostok and Voskhod."

Berry, C. A., "Medical Experience in the Apollo Manned Spaceflights," *Aerospace Medicine* 41(5): 500, 1970.

Bibliography

Berry, C. A.; Coons, D. D.; Catterson, A. D.; and Kelly, G. F., *Gemini Mid-program Conference: Part 1*. Houston, Texas: NASA Manned Spacecraft Center, 1966.

Biological Clocks. Symposium on Quantitative Biology. Vol. 25. Cold Spring Harbor, New York: Long Island Biological Association, 1960. See especially F. Halberg, "Temporal Coordination on Physiologic Functions."

Brown, F. A., Jr. "Living Clocks," *Science* 130: 1535-1544, 1959.

Buenning, E. *Die physiologische Uhr* [The Physiological Clock]. Berlin-Goettingen-Heidelberg: Springer-Verlag, 1958.

Cannon, W. B. *The Wisdom of the Body*. New York: W. W. Norton, 1932.

de Rudder, B. *Uber Sogenannte Kosmische Rhythmen Beim Menschen* [About So-called Cosmic Rhythms in Man]. Leipzig: George Thieme Verlag, 1941.

Enders, G., and Von Frey, M. *"Ueber Schlaftiefe und Schlafmenge"* [Sleep-depth and Sleep Quantity], *Zeitschrift fur Biologie* 90: 79-80, 1930.

Foulkes, David. *The Psychology of Sleep*. New York: Charles Scribner's Sons, 1966.

Gilruth, R. R. "Manned Space Stations." In *Fourth International Symposium on Bioastronautics and the Exploration of Space*. Charles R. Roadman and Hubertus Strughold, eds. Brooks Air Force Base, Texas: Aerospace Medical Division, 1968.

Goltra, E. R. "Time Dilation and Astronaut." In Kenneth F. Gantz, ed. *Man in Space*. New York: Duell, Sloan and Pearce, 1959.

Halberg F. "Circadian (about Twenty-four Hour) Rhythms in Experimental Medicine," *Proceedings of the Royal Society of Medicine*, vol. 56, no. 4, 1963.

———. "Physiologic Rhythms." In *Physiological Problems in Space Exploration*, J.D. Hardy, ed. Springfield, Ill.: Charles C Thomas, 1964.

Harker, Janet E. *The Physiology of Diurnal Rhythms.* London: Cambridge University Press, 1964.

Hastings, J. W. "Unicellular Clocks," *Annual Review of Microbiology,* 13: 297-312, 1959.

Hauty, G. T. "Psychological Problems of Space Flight." In *Physics and Medicine of the Atmosphere and Space.* O. O. Benson and Hubertus Strughold, eds. New York: John Wiley & Sons, Inc., 1960.

Jouvet, M. "The States of Sleep," *Scientific American* 216 (2), 1967.

Juin, G. "Timetable Shifts," *Commercial Pilot,* no. 32. Paris, 1963.

Kleitman, Nathaniel. *Sleep and Wakefulness.* Rev. ed. Chicago: University of Chicago Press, 1963.

Lambertsen, C. J., ed. *Third Symposium on Underwater Physiology. Proceedings.* Baltimore: Williams & Wilkins, 1967.

Luce, Gay G., and Segal, Julius. *Sleep.* New York: Coward-McCann, 1966.

McKenzie, R. E.; Hartman, B. O.; and Welch, B. E. "Observations in the SAM Two-Man Space Cabin Simulator," *Aerospace Medicine* 32: 583-615, 1961.

Mandrovky, B. V. "Soyuz 9 Flight, A Manned Biomedical Mission," *Aerospace Medicine,* vol. 42, no. 2, 1971.

Marti-Ibanez, F. "Health and Travel." In *MD International Symposia,* no. 1. New York: MD Publications, Inc., 1956.

Oswald, Ian. *Sleep.* Baltimore: Penguin Books, 1966.

Parin, V. V.; Volynkin, Y. M.; and Vassilyev, P. V. "Manned Space Flight." COSPAR Symposium, Florence, Italy, 1964.

Pittendrich, C. S. "Perspective in the Study of Biological Clocks." In *Perspective in Marine Biology,* Adriano A. Buzzati-Traverso, ed. Berkeley, Calif.: University of California Press, 1958.

Post, Wiley, and Gatty, Harold. *Around the World in Eight Days: The Flight of the Winnie Mae.* Chicago: Rand McNally, 1931.

Bibliography

Richter, C. P. *Biological Clocks in Medicine and Psychiatry.* Springfield, Ill.: Charles C Thomas, 1965.

Siegel, P. V.; Gerathewoshl, S. J.; and Mohler, S. R. "Time-Zone Effects," *Science* 164: 1249-1255, 1969.

Sollberger, A. *Studies of Temporal Variations in Biological Variables.* Supplement to the report from the Fifth Conference of the Society for Biological Rhythm. Stockholm, 1960.

Steinhart, P. "Der Schlaf des Pferdes" [The Sleep of the Horse]; *Zeitschrift fur Veterinaeren,* 1939.

Steinkamp, G. R.; Hawkins, W. R.; Hauty, G. T.; Burwell, R. R.; and Ward, G. E. *Human Experimentation in the Space Cabin Simulator.* Brooks Air Force Base, Texas: USAF School of Aerospace Medicine, 1959.

Strickland, B. A. "Air-crew Maintenance." In *Aerospace Medicine.* Harry G. Armstrong, ed. Baltimore: Williams & Wilkins, 1960.

Strughold, Hubertus. "Physiological Day-Night Cycle after Global Flight," *Journal of Aviation Medicine* 23: 464-473, 1952.

———. The Physiological Clock in Aeronautics and Astronautics." *Annals of the New York Academy of Sciences* 134: 413-422, 1965.

———. "Circadian Rhythms: Aerospace Medical Aspects." In *Aerospace Medicine.* Hugh W. Randel, ed. Baltimore: Williams & Wilkins, 1970.

Takahashi, Y.; Virginis, D. M.; and Daughaday, W. H. "Growth Hormone Secretion during Sleep," *The Journal of Clinical Investigation,* 47: 2079-2089, 1968.

Titov, Gherman, and Caidin, Martin. *I am Eagle!* Indianapolis: Bobbs-Merrill, 1962.

Verne, Jules. *Le Tour du Monde en Quatre-vingt Jours.* 1873. English translation, *Around the World in Eighty Days.* New York: Charles Scribner's Sons, 1906.

Von Braun, Wernher. *Space Frontier.* New York: Holt, Rinehart and Winston, 1967.

Zim, Herbert S., and Gabrielson, Ira N. *Birds.* New York: Simon and Schuster, 1949.

Index

About the Author

HUBERTUS STRUGHOLD was born in Germany and became a naturalized citizen of the United States in 1956. He received his Ph.D. degree from the University of Muenster in 1922 and his M.D. degree from the University of Wuerzburg in 1923.

After the Second World War, Dr. Strughold became Professor of Physiology and Director of the Physiological Institute at the University of Heidelberg. In 1947 he joined the staff of the United States Air Force School of Aviation Medicine at Randolph Air Force Base, Texas. When the department of Space Medicine was created at the School in 1949, Dr. Strughold was placed in charge of it.

In 1951, the Air University, which included under its command all the educational functions of the Air Force, conferred on Dr. Strughold the academic title Professor of Aviation Medicine and in 1958 the title of Professor of Space Medicine. He is the only person who has been so honored and is often referred to as "The Father of Space Medicine."

From 1957 to 1962, Dr. Strughold held the position of Advisor for Research at the School of Aviation Medicine and the Aerospace Medical Center, Brooks Air Force Base, Texas. In 1960 he was assigned the additional duty of Chairman of the Advanced Studies Group at the Aerospace Medicine Center. In 1962, when the Air Force Systems Command organized the Aerospace Medical Division, Dr. Strughold became Chief Scientist. He retired in 1968 and is now honorary consultant to the Aerospace Medical Division at Brooks Air Force Base. He has been the recipient of many awards for his contributions in space medicine.